Evidence
of
Fire

Evidence
of
Fire

poems & stories

Jennifer Maloney

Clare Songbirds
Publishing House

Clare Songbirds Publishing House Poetry Series
ISBN 978-1-957221-10-6
Clare Songbirds Publishing House
Evidence of Fire © 2023 Jennifer Maloney

cover design by Laura Williams French
cover photos courtesy of
Pete Linforth/the digital artist on Pixabay

Printed in the United States of America
FIRST EDITION

140 Cottage Street
Auburn, New York 13021
www.claresongbirdspub.com

For Christine

Acknowledgements

I wish to thank the following publications for originally publishing the works in this volume.

South Broadway Ghost Society, "A City Story" (April 2023)

Synkroniciti Magazine, "Wild Horses*"* (March 2023), "After the Tower" and "Thee and Me" (September 2023), "Of Goats and Eagles, Frogs and Machines" (November 2023)

Litro Magazine, "How to Catch an Unusual Fish" (February 2023)

My great thanks to Karen Faris, Rachael Ikins, and Bart White for their help editing *Evidence of Fire: Poems and Stories*, and to Barbara Sweeney Evert for her technical expertise.

Contents

Some Notes on How to Survive in Hell

It isn't easy, but it can be done.

First—whatever they feed you, hide under your tongue. Spit it out when the Devil turns away. Just like in a fairy tale, if you eat the food in Hell, you are lost forever.

Do as you're told and expect the consequences. You are doing it wrong. No matter what they said, how they showed you. You're wrong. You have to pay for that. Don't fight. Close your eyes and relax.

Remember—the Devil will be nice sometimes. Will say something kind. You will want to believe it. You will learn that you can't, and learning it will cost you, too.

A magical thing happens. Outside of Hell, you start to make friends. You know how; you have to please the Devil each night. Smile. Crack a sarcastic joke. You're in.

You can have the celery that comes with the wings. No blue cheese! No pizza, egg rolls, ice cream. If you swallow something without thinking, get rid of it—you know how. Food is for the undamned.

You can have a coke. One day, one of the boys pours rum in it, and you make another friend.

Problems will arise.

The nice English teacher wants 500 hundred words on *something your family likes to do together.* How do you respond?

Begin *My father enjoys.* Backspace it away. Write *My mother sometimes likes to.* Stop. Next class, say you forgot. Your teacher will give you another day. Go to the girls' room to puke.

What comes up is as thin and yellow as the dog's vomit. Your classmates giggle on the other side of the stall doors. When you emerge to brush your teeth and hair someone stage whispers *Look, its ANN-orexia! Is she full of BULL-emia? Not anymore!* to a chorus of laughter.

In a couple more hours, you can have your rum and coke. It's all you need. It will be enough.

Try stealing one of the Devil's razor blades. Hide it in your car. At lunch, it's there waiting for you instead of a sandwich. Choose the upper, inner thigh; it's nearly invisible. Make a slit and climb inside. Now you are nearly invisible, too. Head back to class with the thermos of coffee you packed. The teachers think you're very grown-up.

(If you get blood on your pants, act embarrassed. Say you have your period. You don't—you haven't had a period in months. You are reverting to babyhood. You are becoming pure again. The Devil will be pleased.)

Watch the calendar for June. For graduation. Your friends will invite you to their parties. You can't go—the Devil thinks you'll get drunk and have sex at a party someone's mother planned. The devil seems to be deluded regarding where those things might actually occur.

In August, pack the car. Tell the Devil you don't need help moving into the dorm. Say you'll be home at Thanksgiving. The Devil will hug and kiss you and tell you that they're proud of you. Relax. Close your eyes and smile. You know how.

On the outskirts of Hell, just before you get on the highway, there's a little ice cream parlor that you've been to once or twice. Decide to finally get a scoop. You made it! You deserve it. Order vanilla with chocolate jimmies in a cardboard dish.

Get on the highway. Put Tom Petty on loud, as loud as possible. Open the windows and sing *Oh, yeah. All right. Take it easy, baby, make it last all night (make it last all night!)* because you are an American Girl, raised on promises, in Hell.

Let the ice cream melt, untouched, on the seat beside you. You made it out of Hell, but it's still so strangely hot.

Throw the soupy, speckled remains in the trash at the nearest rest stop and find a bathroom. You feel sick.

On the way back to the car, the Travel Plaza Liquor Store

rises before you, the door swinging open like the gates of heaven. Enter prayerfully, with gratitude for your excellent fake ID. Here, in this place, you are at last certain of salvation.

Thee and Me
—after 'Borges and I,' by Jorge Luis Borges

Hand in hand down the sidewalk,
we toddle like preschoolers,
keep track of each other
during fire drills
that aren't really practice.
All these ambulances are real.

I will forgive you for saving me
when you stop taping bandages
over my mouth.
That isn't where I'm bleeding.

I could watch you work a room all day.
Bones on display, a trace of blue,
banging pulse the only clue,
but who sees? Only me, floating
over my own corpse, and of course,
I don't count
(*only her audience knows for sure*), still,
I watch.
I learn.
Nothing is more real
than what we wish to be true.

Why all this interest in my survival,
friend? God knows you don't need me,
the pink elephant in the room. Junkie eyes

case the joint, broken nails open sores,
my last shower a mystery,
my next, not soon enough,
the future an 8-ball,
ask again later.

We aren't joined at the hip.

You aren't locked to my ankle,
a monitor that betrays me
when my mind starts to wander.

Let me go, I will grant you three wishes.
Release me and the ransom shall be paid.

My father is a kingpin,
I am a pearl of great price,
princes climb my hair
like lice, drink my blood
like vampires, and yes—
I invited them in. I will tell you their names

if you lean in close, closer,
get closer,
my dear.

Hand in hand, we are children,
Hansel and Gretel,
the crows eat the breadcrumbs,
the witch walks between us.
She asks us her riddles,
we answer with spellcraft,
we carry our mother's curse in our bones—

a birthmark, an omen, a sign, a foretelling—
you should listen, my sister,
you should heed the old ways:

the river's for drowning,
the grave is a blessing,
the woods are for losing
inconvenient kids—
whose mothers don't want them,
whose fathers can't save them—
the grave is a blessing:

one hungry mouth, fed.

Evelyn McHale's Theory of Flight

Tell my father, I have too many of my mother's tendencies. —
From Evelyn McHale's (AKA 'the most beautiful suicide')
suicide note

Evelyn stands at the edge of the 86th floor,
fixing her makeup. She casts
a practiced eye
at her untrainable lips, her
unsettled mouth, tries
to tame them, once again,
with a slick of cherry red.
She knows, of course,
that it's in the nature of things
(and some people),
to loosen. To unravel
or break or attempt

freedom

from life as a useful tool.
She knows, too,

that it is her job to fix these tendencies.
She fails at it endlessly,
just like her mother. No amount

of paint, powder,
or flattening her cowlick
with foam tape from the hairdresser
will straighten her out. She is untidy.
Loose.

Evelyn's pocketbook loops, not un-prettily,
over a slender wrist, careless of its contents—

a pocket comb, a pair of gloves,
two crow's feathers,
one folded note—
unclasped, trembling slightly,
threatening to spill. Evelyn

is thinking of other things,
as usual. The blue of the excelsior flag
that streams against
the brighter blue of the New York sky.
The ring of its tether against its pole,
like a warning bell,
an emergency signal

help
help
let me go
let me fly

her mind wandering
the way her mother used to wander
from the house, in a bathrobe
she'd worn for days, untied, loose,

her hair streaming behind her,
full of leaf litter, feathers,
burdock, dirt,

slipping each lock,
sliding beneath the gate,
wouldn't be held, left nothing
but untied ribbons
and broken leashes.

There are more than two choices,
there have to be,
Evelyn thinks
as she notices her open-mouthed handbag,
rummages in it, then replaces
her compact, her lipstick,
and clasps it closed,
sets it down neatly
next to her folded coat. There is more
than just tidiness
or disarray. Just staying or leaving. Just death

or a cage, which is death anyway.
She hopes her theory is right, steps into the air,
holding tight
to the feathers she found knotted in her mother's hair.

St. Anthony Searches for Candy and Bones

1.

Tony

St. Anthony's Day Parade   I am 6   7   8   the clowns toss hard candy   tootsie rolls   off the back of an old hook and ladder we kids   scatter into the street   scoop up dum dums jawbreakers boxes of red hots—now

the fireworks. The blanket is scratchy   but warm   the fringe tickles   my sandaled toes   my mother drinks coffee from a thermos   my father is holding the camera

it hurts.   It hurts   the fireworks too loud   I cover my ears try not to cry   my brother balls up his fist   punches my shoulder *you better not cry baby   if they make us go home I'll beat you up*   I believe him   my chipped front tooth   proof   he's not kidding   but it hurts   it hurts   so I decide   I will become the sky   explode   like the night explodes   maybe it will stop—

next day. I close my eyes and see the red worms of my blood vessels.   the sun's a dripping spot   of chrome yellow paint the superballs   are psychedelic   small   hard knots   my brother raises knots   on my forehead   goose eggs   on the back   *you better not cry baby   you better not tell*   but it hurts   it hurts   I never tell but my mother's   coffee attenuated nerves pick up on everything.   She's hypervigilant   hallmark of an abused child   she's neurotically caffeinated   pulls me   into her dark   embrace   says *tell me   tell me   you'll get in trouble if you don't* but I am watching fireworks   behind my eyes   I am exploding   it hurts   it hurts

2.

Tony

Then grown-up how I loved that neighborhood. It's sea-smell it's popcorn and beer ambience   the end of the be-shitted pier

lapped   and licked at   by the big lake   lovers stroll   and
smoke   the heat   like a wet kiss   the breeze up my skirt   flirts
strokes

when we walked back the streetlights were beginning to bloom
like finished dandelions   filaments   of light   detaching
setting themselves afloat on the air   and the neon on the strip
sparkled   flashed its teeth   called to us like a carnival barker
the things it promised   beyond anything, anyone's ability to
give   but what did I know   I only knew the lessons I'd already
learned   and had to keep learning   again and again   there is
nothing there

but sour breath and stretched red smiles.   the clowns are riding
the fire truck again   throwing out prizes and we scurry after
my brother   my lovers   most of all me   I want what's been
promised   what sparkles   what hurts

3.

come around

In my hometown the funeral director stuffed bodies with PVC
pipe   and bags of sawdust   sold the bones   the corneas   until
someone's nephew   noticed his aunt   looking worse in the
coffin than she had at the hospital at the end   empty   sunk-in
like the collapsed roof

of the limousine that cradled the Empire State Building's most
famous suicide   that beauty   landing in repose   ankles crossed
shoes still buckled   somehow   around those ankles   delicate as
wrists in bracelets   her fiancé said she seemed all right to him
her parents said she was a quiet girl   her suicide note said
please forget about me   the most famous jumper in America
resting   on her bier of crumpled metal   the car thank-god
empty   empty as Aunt Jane's burgled chest   her bones going
for $89.95 on Amazon   her final chapter   blank

find what's lost that can't be found

A City Story

Once upon a time, our town owned a story. —William Stafford,
"The Strange Face on the Sand"

This town once told a story.
It was all about our goodness,
our presbyterian Jesus, embodiment
of meek and mild, knew just when
to shut his mouth.

We might've owned the world, but we knew
we owned this city—it looked like us,
grey-faced, combed-over,
bespectacled,
be-cocked.
Our uniforms—

blue coats,
white coats,
topcoats,
coveralls,

badges,
peaked caps,
clipboards
and stethoscopes—

they could have stood up empty,
could have stood up on their own,
so upright were we, so stiff,
so *erect* with straightness—
the bleach of it burning
our eyes, our throats,
our thoughts—our thoughts

were all about this city,
what it needed,
what we'd give it,
whether it needed it or not:

white-gloved crossing guards
blonde, baton'd majorettes,
a thousand brushcut lunchpails,
a parade of white bread wonder
fed into the factory daily—
while we kept

the wheels turning,
kept the peace
at the business end of the nightstick,
kept the hysterical sedated
with TV and Black Velvet
and small pills
for big-mouthed women—

this town once had a story,
a secret underneath its skirt—
the pressure point of the club handshake,
the sweet grease for the palm-reader—
the future
was ready-to-wear.
We believed it, believed in it, believed we'd

get
what we wanted,
the trophies we paid for,
the money, the manna, the mammon—
we'd get everything
we deserved.

It's not the dogs,
not the fire hoses
that ended this tale.

It's the photographs the press took,
how it looked
on the news. Operations interrupted
for awhile as we smiled,
shook our heads, said
what a shame,
we must do better…and we got better.

At the story.

At the inside jokes.
Got degrees in Women's Studies, hid
in Diversity Departments.

Learned to murder Black kids,
but phrase it right on resumés,
get a job as the director
of the Police Accountability Board.

This story keeps on rolling.
This story is a running joke.

This town elects its drug dealers,
pays its whores with plummy titles,
keeps its finger on the pulse,
says we have no DNR,

so the ventilator breathes for us,
the psyche meds think
and dream for us,
the generic Viagra fucks for us,
the Trazodone tucks us in.

In fiction, there are endings,
there is meaning, sometimes lessons,
but this story, like this city,
has a life all its own.

And who am I to judge it?
To defend it? To defund it?
Who am I to count its blessings?
Or to number all its bones?

This city is American.
This city could be anywhere.
This city never pays for guardrails
if it can vote for guns—

this city is my hometown. This city isn't getting better.
This city has no place for me.

It's my hometown,
but it's not
home.

Like Sylvia

Being born a woman is my awful tragedy. —Sylvia Plath

I wanted only freedom.
To ride my thumb
across this country. I wanted

a year. Maybe two, maybe

(my whole life) more

to talk to strangers
in the cabs of semis, to watch
America
speed past my elbow. I could feel

entire lives
being lived in the driver's seats I imagined
beside me:

old men that munched pistachios,
slow-piloting Impalas
that stank of pine freshener
and Swisher Sweets,

Native families in windowless
white cargo vans, solemn babies,
hand-rolled cigarettes
offered from a plastic baggie,

realtors in coral lipstick
and bright blond bangs,
who'd pick me up
because I looked cold. At eighteen,
I could see a slender rind
of chance, gleaming

at the western horizon
(orange and fragrant
as a cantaloupe

in the jouncing back of an open bed pick-up,
liquid, and already leaking
through my grasp),
a wedge of light, a chisel, perhaps,
levering open a locked door. A crack,
just high and wide enough
to limbo underneath.
I thought I'd make it. It was right there,

at the terminus
of the first digit
of my right hand, jerked
and pointed anywhere

(take me anywhere),

I only wanted to be anywhere

but here, constantly reminded
of my constant companions:

these tits.
This blood.
These cracks—
that I could only fall through,

never dance beneath.

Help Wanted

Every time I sneeze,
the devil blesses *me*,
a little 'thank-you',
because I give his demons work—

each one a shiftless bastard, lazy,
inclined to drinking on the job,
pointing fingers when they lose them
in machinery—

they need purpose and direction.
This business manufactures sin.

Has several vital openings
to be filled: now hiring

liars
for the assembly line
to fit the pieces into place,
engineers to tweak the measurements,
make sure the stories
hang together,

agile thinkers in the boardroom
to beef up the bottom line,
cut out all the deadwood
(a few fingers at a time),

we need salesmen
on the showroom floor,
with shiny teeth that slice
through thickets
of paperwork,
make it less work for the buyer—we want

gophers
and assistants,
chair-fillers, talking heads,
heart-attacks-waiting-to-happen,

paper pushers.
Middle management.
Corporate raiders. Golden parachutes.
We need loudmouths in loud ties,
and quiet-quitters.

There's room at the top,
room in the middle,
for headhunters in personnel,
number-crunchers in accounting—
they're counting coup, taking scalps,
in the offices of our lawyers,
all security is on the take:
professional industrial spies.

We stay open.
24/7.
No experience required.
All education levels needed,
all shifts hiring.

An Equal Opportunity Employer,
we offer a living wage—
if just a living's what you're after.
We *could* set you up much better.

See me in my office,
after hours.

After the Tower

My promises go out at night and forget themselves. They wear short skirts and body glitter. They whistle at passing cars.

They get sloppy drunk and break, just like the plate glass windows that front the Cozee Inn Tavern on Main, and scatter in sharp, shiny puddles on its scuffed linoleum tiles. They don't think about security cameras as they stand sobbing on the sidewalk, screaming *you bastard, you sonofabitch, I loved you.*

My promises make bargains with themselves. *This time, just one. The baby will be fine, she's sleeping.* My promises believe their own bullshit.

Sometimes they crouch on my tongue, feeling for vibrations the way a trapdoor spider waits for prey. Sometimes they fall out of my mouth and I can't stop them, like roses, or diamonds, or venomous toads.

You feel them slip beneath your skin like a needle. It used to hurt. Now you crave the prick as much as the poison.

My promises pass out in alleyways and wake up in jail cells. They wish they were dead, pray for it. It can't come soon enough. It's going to be here sooner than they think.

You love my promises. They climb into bed with you and we all close our eyes for a little while. You hold them, comfort them, and they forgive you everything. When we wake up, my promises have become yours.

Someday my promises are going to grow up, just like the baby. They will sling a duffle bag over their shoulder, buy a bus ticket, never look back. Out there in the wide world, there are people who keep their promises, people I've never met, but in whose existence, I believe, like Santa Claus and Jesus. My promises will run away and find their real mother, a princess in a tower, from whom they were stolen long ago.

Alien

In the bathroom I present
my closing arguments.
The mirror says

I'm smooth,
I'm cool, says
You cannot fuck with me,
reminds me,
I can stop shaking
anytime I want.

To the mirror
I repeat my affirmations:
I am beautiful. I deserve love.
I have boundaries today.

There are evenings
when the moon simply *sprawls*
at the horizon. Fat, yellow,
benign as a Buddha, but tonight,

she's white-knuckling it
like a three-day-dry drunk,
wrapped tight and riding high,
her glaring skullcap eye
just jonesin' for a fight.

The girl in the bathroom mirror
thinks she's already skinny enough.
Only works out because she likes it,
never considers
throwing up before she weighs herself,
because she doesn't weigh herself.
She's beautiful.
She is loved.

I walk among these humans,
show them this face I've borrowed,
pocked and pale.

Just as though I'm one of them.
They never seem to guess the truth.

And the girl in the mirror,
with her smooth iron eyes,
smiles right at them,
and does not betray me.

Lost Angel

Tom deployed,
in the last four years of Viet Nam,
aboard an aircraft carrier,
the Nimitz.
He fixed fighter jets.
Combat operations

basically over, life aboard ship
was pretty smooth:
boy scout camp for grown-ups,
he called it,
three hots and a cot.

One time, all fucked-up
on hasheesh and windowpane,
he released his A-6 Intruder and started
below decks, a hundred pounds
of tie-down chains
swinging heavy from his shoulders.
He was 25 feet

from the nose of his plane,
headed for the line shack
and a cup of hot coffee,
when the A-7 Corsair,
perched in the catapult opposite,
lurched
and let go.
He hit the deck.

Lost the chains, but kept his head,
the plane missing it by inches.
His mother said
he must have had an angel on his shoulder.
He just laughed.
What else can you do?

It's that or cry,
and he wouldn't cry, not then,

and not
when bent beneath his father's belt,
silver buckle
swinging heavy
in the oil-stink,
the frozen air
of the dim garage on Pierpont Street,
one ear still red and ringing
from the Good Sister's slap. Tommy,

focused on the sunbeams
painting flaming swords on concrete,
promised, someday, Daddy,
I'll be big enough to fight, hit back.
Til then, he laughed!

Fuck you, old man, fuck you,
I'll piss on your grave!
while his beer-fat father
bellowed himself into a coronary.

Where was Tommy's
Guardian Angel then?
Having a smoke?
Waxing his feathers?

There was a time he'd thought
Maybe I could be a rock star.
He learned guitar pretty fast
while tripping, was a pretty boy,
had pretty girls.
Instead,

he knocked me up while home on leave,
and all those dreams
went sliding down my thighs.

Thirty years later, off the factory floor,
all the babies grown and gone,
his long black curls now steel wire,
his fingers slower on the frets,

he's retired,
eating cheeseburgers for free, with me,
on Veteran's Day,
and the kids in paper hats all shake his hand, say
Thank you for your service,

while once again
he tells me that the Navy wasn't more
than boy scout camp,
three hots and a cot
and some pretty good drugs.

It wasn't service.
He was freed.

And hey—his Guardian Angel finally found him,
after all those years stumbling blind
around Rochester garages;

sending up smoke signals,
waving his sword,
even dropping a trail
of glossy feathers behind him,
in hopes the boy would follow.

The Moon is Just Another Rock

I carry your voice around in my pocket.
I don't listen, but I don't delete.
The way of flesh is absorption.

You're just a small secret:
an old pipe I never threw away,
the needle still stashed
behind a tile in the bathroom.
You're what I squeeze when I feel stressed,
the secret drawer in the credenza,
the Xanax I dropped
in the martini. You're for emergencies.
You're just-in-case. You
are insurance.

The way of flesh is absorption.

Leave a hundred coded notes, words won't
tear open in my throat, a piñata stuffed
with ciphers and desire. I am not listening.
I am a stone.
My body is translucent,

I am as pale as the moon,
white as the flamingo
that only reddens with the krill,
their feathers pure before pollution.
The way of flesh is absorption, but

you're just my little secret.
A worry doll, a rosary, a fetish I can rub.
You are nothing.

A little power I can hold.
A little knowledge, unabsorbed,
I do not hear you.
You cannot hurt me.
I am not flesh,

I am the pockmarked face of the moon.

The Bride

I am a soft creature
in the hands of a god—a thing
with the head of a bear, raptor's talons.

Around me, I know,
the world tumbles forward.
They smell my corruption,
despise me.

What city is this, what necropolis?
The signs are all ciphers,
the map only shimmers.

Oh, Little Death, trembling
on the altar beside me,
if I give you my power,
will you give me your juice?
You're an animal like me,
but with much sharper teeth—
can I trust you? Will you be my friend?

Here is my sweet belly.
Give me your tender neck,
close your eyes, take my hand,
as the god bellows need
(we must wait for judgement,
we must hope for mercy),
will we stagger together
from the doors of the temple,
sacrificed, sated, and stupid with joy?
You and I, Little Death,

bloody and blessed,
bearing collars of jewel-toned bruises—
tattoos in the shape of a kiss—

branded and tamed,
owned and named,
made domestic
with beatings and bliss?

Of Goats and Eagles, Frogs and Machines

We think we are funicular,
little goats on beveled hooves,
that we can balance
as we climb,
teeter through the kitchen
in our kitten heels,
bear breakfasts on our breasts,
carry coffee, serve

thanks-chris-easter dinner,
and believe we can

fly away

using only the marabou feathers
stitched to our negligees.

My second baby.
So red, so black,
the queen of hearts.

Her body a frog,
knees splayed
in the bath, arms flung out
in startle. Pink pinprick
nipples, the shriveling husk
of umbilicus, I stand above
her eyes, her mouth

searching, rooting, wanting,
and say out loud
I hate you,
and in the silent room I think,

I can leave.
I can just leave.

I am twenty-four.
I drag exhaustion behind me
like a dirty blanket,
my only desire
to curl beneath it like a cat.

But I am a portable milk machine.

I'm the only person in the room
who can smell a dirty diaper.

Mother says she'll help me,
but if I sleep when the baby sleeps,
I can't keep an eye on Mother, too.

I am looking at the baby,
little frog in the bath.
She is red and she is black,
the queen of hearts.

She looks exactly like her father.
She *wants*.
Exactly like her father—

an open, working mouth
that I must feed—my body screams
to be consumed,
requires me to be required.

This thing wants only all of me,
that's all.

I thought
I was a mountain goat,
sturdy, light, easy-footed.
I thought,
maybe these feathers
meant I was an eagle, but

I am a milk machine.
Everything inside me is just food.
I am mother
to a frog, so red, so black,
floating in the bath.

I can leave.

I can walk away.

Can't a frog
swim?

Almost Roxanne

What was your almost-name? My mother says I was almost
Emily—seen on Facebook

She was almost Roxanne, almost
finished drama school,
would have graduated
in the same class as a now-famous Latino actor.
Was his scene partner, almost dated him, but

almost Roxanne almost
didn't make it through the first semester.
She almost killed herself
with wine
and whippets, so, instead
of almost winning

a Tony, an Oscar,
she went home
to waitress at the Big Boy,
learned how to act
happy
over slices of mile-high strawberry pie.

In 1982, Big Boy stood proudly
in red-checked overalls,
his chocolate forelock,
a triumph of concrete casting,
waved at the passing cars
out on West Ridge Road,
on almost the very same spot

where West Winds Wine, Liquors,
and Spirits stands now, adjacent
to the Friendly Motel,
where she almost always rents

the room at the far end,
so she has to at least
think about it,

before deciding
to stumble back across the parking lot
to replace this bottle of Old Times
that's almost empty,
has to be pretty certain
she won't sleep without it,
almost positive.
The Friendly Motel

boasts ten small cabins,
side-by-side, crummy little bathrooms
at the back, rust-streaked sinks,
torn shower curtains,
dusty, wrapped soaps almost never used.
You can tell it's friendly

by the way
the skinny chain of rooms
almost seems to crook
a come-hither finger
at the bustling traffic,
to wave at passing cars
like Big Boy used to:

Come on in! It's time,
just like Old Times, you have
almost enough time
to get mile-high again.

Almost Roxanne almost got married,
twice, almost had a baby,
had a miscarriage instead.

Isn't that a funny way
to think about having something?
Having the thing
that means you have nothing?
Having a miscarriage.
It's like picking up a handful of dirt
and saying
now I own real estate.
Anyway,

she almost had everything she ever wanted,
a couple of different times,
ended up instead
with a little addiction
and a spontaneous abortion,
but now she has a new bottle
of Old Times,
almost full.

She's almost forgotten
about the Latino actor,
grinning mustache-ily
from Pepsi commercials,
forgotten the way he said her name,
his eyes, almost the same
warm chocolate
as Big Boy's beckoning bangs,
nearly as friendly
as this crappy motel. *Her* name,
not who she almost was, no,
but the girl in 1981, the one

who's name sounded almost as if
it began with a Y, a friendly sound,
like it could melt between his lips,
might dissolve
beneath the pressure
of his tongue,
almost as sweet
as strawberry pie.

Instinct

In this dream, fur
that has drowsed along my spine for years
awakes. Lunges. Growls. I am dragged—

I am racing down a path—

I am following the map to all my shame—

my fur is a wolf pack running down an elk,
shot, bleeding. My fur,

its deep-belly hunger,
its glassy bones that sing beneath my skin—
hum   vibrate   rattle   I am clinging to my fur,
I grip it

and it barrels through the night, trees and snow
a blur   a fleeing gleam   under moonlight

and my companions run with me,
know the way. My fur

is focused on the hunt. Blood steams
in its nostrils, water streams
from its mouth. This chase—

muscles coil, spring
(the body refuses no demands),
shapeshifts, flies and falls,
instinct pounding   in its chest
ears   loins—

my fur closes on its prey. The pack slows
and circles.

The elk is spent.

Stumbles. Sways   graceless   heavy

knees crumple into snow

my fur is on the beast in seconds. The pack feasts.
In this dream, my fur

knows the path to all my shame. Follows
where the blood leads. Feeds.

In the morning,
I find that I have stained my sheets.
My fur would know what happens next, but

I am the elk.

Wild Horses

At the cabin I am waiting at the door for you
to tell me again, as you always must.

You climb out of the coffin
of your bile-green Chevrolet,
frost-blossoms fading
from its windshield as it warms,
and you stalk back
up the path
to tell me what you think
I'd better know, what I
really should be told
about myself
before you finally,
finally,
get on with it
and leave.

Behind me,
the woodstove glows
red and black
through its little barred window, so hot
that sweat
bumps down the knobs
of my spine, collects
in the dimple that sits
just above the cleft of my buttocks,
like a little cup of fear.

As I stand and wait for you
to tell me more about myself, to teach me,
again, the lesson held
in the pointed finger, the raised palm,
I see

the little carved horses
that guard the cabin stairs
rear up, snort,
mimic the smoke

and stink
of the Chevy's exhaust
and the gut-sour breath
that steams from your mouth,
and one wild, stone eye
rolls back in a carved socket, as you

and your scuffed leather boots
and the swinging slabs of your arms
march toward the three of us—

and the dawn is flush
with color, crimson and purple
as the booze-blooms in your cheeks, strengthening
pinks, lavenders, a simmering
blood-orange sun begins
its slow slide up
the edge of the world, I see it all
as you come for me, for me
and the little horses
that paw the air
and whinny—

remember when we bought this place?

We stood together on the front porch, your uncurled fist
resting on the small of my back, just above
the cleft of my buttocks. We talked

about how we could grow here, leave
the beer and the cocaine and cigarettes behind us.
We talked about the horses.
I'll paint them white, you said, but
you didn't.

Icono-Plastic

He says *I want to eat you up,* cups my breast in his two palms, the body of Christ. I lift my face to heaven, howl *Hallelujah,* scream *Amen.*

$7000 is what it took to get here. $7000, 12 hours on the table, and this narcotized dream: yanked backwards through a tunnel, wide-eyed faces reeling past, as the Great God of Missed Opportunities hooks fingers through my belt-loops, drags me, ass-first, back into the light. His enormous crook has caught me 'round the middle, pulled me off whatever stage in hell on which I've just begun my final burlesque and He delivers me, again, into the velvet wings of life.

It costs $7000 to sob on the toilet of the recovery room while a dream nurse wails *Something's wrong, take her back, we did something wrong!*

He lifts my other breast and tongues the scar beneath it. *Perfect,* he whispers, and my head jingles like a cash register. $1500, at least, for just that one, alone.

The knife marks stretch elbows to armpits. Run beneath both breasts, encircle each nipple. He touches two that slice diagonals beneath my shoulder blades, and I say *That's what's left of my wings.* He laughs at the idea that I could be an angel, although I may be his savior, neither of us certain. Clearly, we are blessed, unabashed in our nakedness, even as the serpent starts to writhe. *Confess,* he whispers, and presses me against the apple tree, anoints me with the sacred oil.

$7000 isn't much to ask for absence. Absolution is costly. It's expensive to be remade in the image of God, but consider the alternative: a name, unloved, chiseled from history, the smashed faces of the statues of kings. The new regime eradicates what power the old may try to grip, even in death. Attempts to stop the haunting.

I know I will haunt no one. Power costs $7000, more or less, and I must wield it while I can.

So, I do as he commands. Confess. Yes, the snake's sweet tongue has whispered me his secrets. I have eaten the fruit, consumed the knowledge of my sin. I found it good, to discover I am just the same as anybody else, equal in flaw and grace. And with my teeth I will tear out the dark heart of the nut, even as the dream insists *Take her back, put her back to sleep. Something must be wrong.*

A Bargain at Any Price

Perfection, of a kind, was what he
* was after,*
and the poetry he invented was
* easy to understand;*
he knew human folly like the back
* of his hand,*
and was greatly interested in armies
* and fleets;*
when he laughed, respectable
* senators burst with laughter,*
and when he cried the little
* children died in the streets.*
        —W.H. Auden, "Epitaph on a Tyrant"

I.
My father introduces us.
My business associate.
Bogat, mouths my father. Rubs fingers
together. *Rich American.*
I understand.

We are in some trouble at home.
Mother's credit declined
at the shops, restaurants
suddenly fully booked,
no tables available.
Gossip
dances in the air,

like the flies
that hover over a corpse,
and people are just animals.
They can smell blood, too.

II.
He takes me to a disco.
I wear a good, knock-off Chanel,
Mother's pearls.
He's in some American designer's
shiny blue uniform,
long red tie.

He knows the proprietors.
He buys cocaine. I laugh

at everything he says,
show him my teeth
when he touches me.

III.
C'mon, you know you want it, he pants
in his slick rental car with too-dark
tinted windows. White powder
sugars the carpet,
a jumble of multi-colored pills
strewn across the console.

Hands.
Fingers.
Mouth.
I squirm

and twist, pretend
to giggle, that it's a game.
I'm on my period, I lie,
and he sits back, disgusted.

Men like him, they think
there is good blood
and bad blood, and the stuff
between a woman's legs
is gross. It makes me laugh.
Blood is blood,
my squeamish friend,
and i have seen it pour

from the necks
of former friends of the regime,
trickle from the mouths,
the still-surprised eyes,
of the freshly hanged.
What comes from me, or any woman,
is clean as Holy Water
compared to that.

IV.
He's leaving in two weeks.
I spin him up every chance I get,

all my clothes low-cut, tight.
I don't do drugs with him,
but I never suggest
that he shouldn't.

I wear a lot of white. Pastels.
I let him touch me.
I touch him, too, but
I won't let him fuck me.
We are religious, I say.
I never explain which one.
*My mama—it would
scandalize.*

V.
He gets his divorce.
We run to the courthouse,
and neither of us
mention religion again.

VI.
Afterward,
my father kisses me.
Dobro opravljen, he whispers.
Well done.

VII.
Fifteen years later, Mama and Papa
die in Palm Beach, of old age,
instead of prison in Novo Mesto.

I have a son.
I tell him, *You and me. We
are what matters. Family.
Only family.*

He doesn't understand
that his father is not family.

His father is what I did
for family.

VIII.
On the evening news,
a young reporter stands
at the edge of a black river
not very far from here.
Small bodies drift, face-down, wheel
in slow pirouettes, like fallen leaves
in a sluggish current. On the riverbank,
laid out like fish, their parents.
Mouths wide, eyes

so surprised,
they gape at the sky,
at God,
the silent stars, the huge,
indifferent moon.
What happened?
they wonder,
the fools. The pawns.
*What just
happened?*

IX.
I don't ask any questions.
Here, in this new land,
in my spacious home,
my bedroom door locked
against the price I paid, I sleep
easily, and my boy
sleeps the sleep of innocents.
Sleeps like the dead.

Costs

Survival is expensive.
Not dying at seventeen
(in handsome Frankie's new Impala,
or maybe in a warm bath
of hydrocodone and red wine),
investing, instead,

in the high-stakes, volatile,
and fast-moving crack cocaine industry,
has cost me

a handful of teeth,
a new hip at 60,
a certain amount of self-respect,
and the love of at least one
of my children. I'd love to know

how much money I actually made
selling blowjobs in SUV's and alleyways
at twenty bucks a pop, tax free,
for nearly four years. I'll bet it's quite a bit,
though still not enough
to finance an American retirement.

Here's an interesting fact: did you know
that moon dust
in the lunar lander
smells like gunpowder,
and no one knows exactly why?
It may be
that the very dry dust reacts
with moisture
in human nasal passages, releases
four billion years
of dangling chemical bonds
that just can't wait to partner up,
like people
who love the scent of danger.

More than half of it is made of glass.
If you inhale too much,
your lungs will shred like party streamers—
but don't worry. You won't be tempted.
Back on earth,

moon dust loses
its exciting smell, and who cares
about some dirty old rocks
if they can't kill you anymore?

How to Catch an Unusual Fish

I pick up my lover and we drive to the abandoned ice cream shop, pull around behind it. It's still daylight, which is good. The cops only bother idling cars at night.

We do what we came to do, with all its noises, heavy breathing, and changes of position, not always easy in my little Honda. Afterwards, I open the glove box for the tissues I keep there. He smiles. *You're so efficient.* Well, it's been five years—three, if you subtract the plague years, but still. By now, I know what I need. What we both need.

The agreement is that we will be quick, but still good to each other. That neither of us is looking for romance. We are adults, not teenagers, and we both have lives—people—that must not be harmed. This is simply—release. Fun. That's the agreement.

He sits quietly for a moment after the clean-up, then takes my hand. Looks at me for a little while. Looks at me until my eyes start to sting and blur and I have to turn my head. "Stop," I whisper. He squeezes my hand. He stops.

We head back to his car, town police none the wiser. It's getting harder to find places to go. He keeps suggesting a hotel. I think it's a bad idea.

We say goodbye. He slides into his seat, picks up his phone. She's probably texted. *Please pick up milk and bread on the way home.*

I have a text from home, too. *How do you catch an unusual fish? Unique up on it.*

An American Retirement

Warm in the car, we sidle up to the drive-thru, bashful and eager as virgins, fantasizing chocolate covered doughnuts, a flap of cheese on a slender patty, cream mellowing hot coffee. We pass money and menu items back and forth like tongues, then slide away from the window's open mouth, breathing into ours like a lover, steaming the glass. Our own window glides up, seals in the scent of breakfast or lunch or dinner, and then we are on our way.

I open your coffee, pour the torn sugar packet, stir. Dissolve it, and my kiss, testing for heat and sweet, before I hand it to you. This ritual a comfort, even as our waists thicken and our hair grays.

We know we ought to bicycle to work. We should eat the apples wrinkling in the basket at home. We should move to a Blue Zone, maybe Arizona, where the old people all live to 101, 102, playing pickleball on new hips and knees. We know these truths to be self-evident—we can see them happening in front of us. We point at them and say *look, see? We should, we ought to*, and then we put some ketchup on the fries and stuff three or four of those lovely, salty fingers in our mouths, or open wide and let the sugar-glaze dissolve the terrible knowledge that we can also watch unfolding before us: our bodies crumbling sweetly into hot coffee, our own new knees settling into the contours of the car's bucket seats as we eat our breakfast, lunch or dinner on the way to the sales job at the mattress store or the temp placement at the GM factory or the call center position or even to work at a different fast food place.

We understand the *shoulds* of our lives but do our best not to think about them. The apples and the Blue Zones. The happy, healthy, pickleball-playing seniors the internet says we could be if we would only try. The distances our fourteen-year-old car with its rotted floor and eternal check engine light will never travel.

We can only afford to drive to work. We can only afford the small pleasures of the already damned. They will have to be enough: every morning, with your sugar, I will drop my kiss into your cup, the secret ingredient to this long, long, very long life.

Evidence of Fire

...And that's when he noticed soot on the ceiling...
—from "This Small-Brained Human Species May Have Buried
Its Dead, Controlled Fire and Made Art," Kate Wong, *Scientific
American Magazine*, June 5, 2023

Unbuckle, unzip, undo, while what is sweet in you
transmutes to salt, preserves these fugitive kisses,
fills my mouth with ocean, stings
where I have bitten my tongue.

I cannot think. My brain
is filled with chemical mud.
I have given over. Surrendered

to the car's self-driving mechanism—
I am dreaming in the backseat,
already in your bed
as I hydroplane, skate
from lane to lane.

I cannot write,
but I can hieroglyph a message,
make the sign for water, for drowning.
I can leave the pattern
of my fingers on your skin,
like handprints in red ochre stamped
to cave walls, or soot on the ceiling
that says *we burned here*, but

we don't even leave each other
voicemails. Delete
texts and emails as they arrive,

and if I lick the letters of my name
between your thighs,
between your sighs,
nobody knows.

We can't save ourselves
with salt or soot. Kiss me anyway,
and hold me even if you can't hold on.
If we are just chemicals, or dreams,
or simply smoke,
then we are just the same as everyone else:
the traces we leave if not invisible,
at least not completely knowable;
all those poor archeologists,
standing over our bones,
scratching their heads, mystified.
You and me,

a genuine wonder.

Jennifer Maloney is a disabled writer living with chronic illness. Find her poetry and fiction in *Litro Magazine*, *Synkroniciti Magazine*, *ImageOutWrite* volumes 7 and 8, and *SHIFT: A Publication of MTSU Write*. She has stories forthcoming in *Fantasy and Science Fiction Magazine* and *Literally Stories*. Jennifer is the co-editor of the poetry anthology *Moving Images: Poetry Inspired by Film* (Before Your Quiet Eyes Publishing, 2021), and the author of *Don't Let God Know You Are Singing*, forthcoming from the same fine press. She is also a parent, a partner, and a very lucky friend. She stays grateful so she can stay sober, and still feels all the things.